Energy Sustainability: Planning, Policy

Dr. Hemant Pathak

Copyright © 2013 Dr. Hemant Pathak

All rights reserved.

ISBN: 1492920436
ISBN-13: 978-1492920434

DEDICATION

Dedicated to Shri Sainath Maharaj the all omnipotent of world the most merciful.

CONTENTS

	Foreword	5
	Glossary	8
1	Introduction	16
2	Sustainable energy	17
3	India's energy policy	20
4	Energy efficiency	25
5	Renewable energy potential	26
6	References	28

Foreword

Energy supply from renewable resources is now essential global strategy, especially when there is responsibility for the environment and for sustainability.

The world is moving towards a sustainable energy future with an emphasis on energy efficiency and use of renewable energy sources. A finite planet cannot support infinitely increasing consumption of resources and hence the motto of present times must be to 3 R principal - "Reduce, Reuse, Recycle".

Energy Sustainability: Planning, Policy; provides a unique insight into the problems our planet faces in terms of clean energy resources proven technical and economic importance worldwide. This books Written for academics, researchers and practitioners working in Energy field, expressed comprehensive and interdisciplinary focus on the current energy demand to enhance energy conservation outcomes.

This book has been provided to be utilized by all people concerned with energy conservation in all the industries in world.

This book provides an essential guide to researchers, it offers: various aspects of Energy Planning and Policy in present scenario.

Simply explained, Energy Sustainability: Planning, Policy is an important book bringing together diverse viewpoints from Industries and state agencies and regulators, for all who wish to make a difference in how to plan and manage our Energy resources.

Dr. Hemant Pathak
M.Sc. (Gold medalist), Ph. D.
Assistant Professor of Engineering Chemistry
Indira Gandhi Govt. Engineering college, Sagar, MP, India

Acronymns

CH_4 methane

CNG compressed natural gas

CO_2 carbon dioxide

CO_2e CO2 equivalent

CPP Critical Peak Pricing

CRIS Climate Registry Information System

ECMP Energy Conservation and Management Plan

EMAP Energy Management Action Plan

EMG Energy Management Group

EMS Environmental Management System

GHG greenhouse gas

GIS Geographic Information System

N2O nitrous oxide

NPV net-present value

O&M operations and maintenance

RECs renewable energy credits

REP Renewable Energy Program (Policy)

Energy Sustainability: Planning, Policy

Energy units and conversion factors

Temperature Kelvin (K)

Commonly used temperature units

Celsius (C), Fahrenheit (F)

0°C = 273.15 K = 32°F   1°F − 5/9°C   1°C − 1 K

Fahrenheit temperature = 1.8 (Celsius temperature) + 32

Derived SI units

Heat: Quantity of heat, work, energy joule (J)

Heat flow rate, power watt (W)

Heat flow rate watt/m2

Thermal conductivity W/mK

Glossary

Abatement — The reduction or elimination of pollution.

Acid rain — The precipitation of dilute solutions of strong mineral acids, formed by the mixing in the atmosphere of various industrial pollutants

Act — A law

Aerosol — Particles of solid or liquid matter than can remain suspended in air from a few minutes to many months depending on the particle size and weight.

Air pollution — Toxic or radioactive gases or particulate matter introduced into the atmosphere, usually as a result of human activity.

Ash — Incombustible residue left over after incineration or other thermal processes.

Bagasse — The fiber residue that remains after juice extraction from sugarcane.

Btu — The abbreviation for British Thermal Unit(s).

Byproduct — A secondary or additional product resulting from the feedstock use of energy or the processing of nonenergy materials. For example, the more common byproducts of coke ovens are coal gas, tar, and a mixture of benzene, toluene, and xylenes (BTX).

Bioenergy — The conversion of biomass into useful forms of

	energy such as heat, electricity and liquid fuels.
Biogas	Gas produced by the biological process of anaerobic (without air) digestion of organic material.
Biomass	Organic, non-fossil material of biological origin constituting an exploitable energy source.
Carbon dioxide (CO_2)	A colorless, odorless, non-poisonous gas that is a normal part of Earth's atmosphere. Carbon dioxide is a product of fossil-fuel combustion as well as other processes. It is considered a greenhouse gas as it traps heat (infrared energy) radiated by the Earth into the atmosphere and thereby contributes to the potential for global warming. The global warming potential (GWP) of other greenhouse gases is measured in relation to that of carbon dioxide, which by international scientific convention is assigned a value of one.
Carbon sink	A reservoir that absorbs or takes up released carbon from another part of the carbon cycle. The four sinks, which are regions of the Earth within which carbon behaves in a systematic manner, are the atmosphere, terrestrial biosphere (usually including freshwater systems), oceans, and sediments (including fossil fuels)
Climate change	A regional change in temperature and weather patterns. Current science indicates a discernible link between climate change over the last century and human activity, specifically the burning

of fossil fuels.

Clean Energy Technologies   Electricity and/or heat producing systems that produce negligible or minimal amounts of environmental pollution compared with conventional technologies.

Coal   A readily combustible black or brownish-black rock whose composition, including inherent moisture, consists of more than 50 percent by weight and more than 70 percent by volume of carbonaceous material. It is formed from plant remains that have been compacted, hardened, chemically altered, and metamorphosed by heat and pressure over geologic time.

Combustion   Burning. Many important pollutants, such as sulfur dioxide, nitrogen oxides, and particulates (PM-10) are combustion products, often products of the burning of fuels such as coal, oil, gas, and wood.

Contamination   The act of polluting or making impure; any indication of chemical, sediment, or biological impurities.

Dust   Solid particulate matter that can become airborne.

Ecosystem   An interactive system that includes the organisms of a natural community association together with their abiotic physical, chemical, and geochemical environment.

Electric current	The flow of electric charge. The preferred unit of measure is the ampere.
Electric energy	The ability of an electric current to produce work, heat, light, or other forms of energy. It is measured in kilowatthours.
Electric generation industry	The electric generation industry includes the "electric power sector" (utility generators and independent power producers) and industrial and commercial power generators, including combined-heat-and-power producers, but excludes units at single-family dwellings.
Emission	Release of pollutants into the air from a source. We say sources emit pollutants. Continuous emission monitoring systems (CEMS) are machines, which some large sources are required to install, to make continuous measurements of pollutant release.
Emissions coefficient	A unique value for scaling emissions to activity data in terms of a standard rate of emissions per unit of activity
Electricity	A form of energy characterized by the presence and motion of elementary charged particles generated by friction, induction, or chemical change.
Energy	The capacity for doing work as measured by the capability of doing work (potential energy) or the conversion of this capability to motion (kinetic

energy).

Energy consumption The use of energy as a source of heat or power or as a raw material input to a manufacturing process.

Energy Audit An assessment of a home's energy use. These include a number of different types of surveys, including (in increasing order of cost and complexity): online audits, in-home home energy surveys, diagnostic home energy surveys, and comprehensive home energy audits.

Energy Conservation Saving energy by doing with less or doing without (e.g., setting thermostats lower in winter and higher in summer; turning off lights; taking shorter showers; turning off air conditioners; etc.).

Energy Efficiency A ratio of service provided to energy input. Services provided can include buildings-sector end uses such as lighting, refrigeration, and heating: industrial processes; or vehicle transportation. Unlike conservation, which involves some reduction of service, energy efficiency provides energy reductions without sacrifice of service.

Energy loss Deleted because there is no need for a general term to encompass all forms of energy loss. Terms referring to losses specific to particular energy sources are defined separately.

Exposure	The concentration of the pollutant in the air multiplied by the population exposed to that concentration over a specified time period.
Fossil fuels	Fuels such as coal, oil, and natural gas; so-called because they are the remains of ancient plant and animal life.
Gasification	Combustible gas called producer-gas produced from biomass through a high temperature thermo-chemical process. Involves burning biomass without sufficient air for full combustion, but with enough air to convert the solid biomass into a gaseous fuel.
Geothermal	Natural heat extracted from the earth's crust using its vertical thermal gradient, where discontinuity in the earth's crust
Global warming	increase in the average temperature of the earth's surface.
Greenhouse gases	Atmospheric gases such as carbon dioxide, methane, chlorofluorocarbons, nitrous oxide, ozone, and water vapor that slow the passage of re-radiated heat through the Earth's atmosphere.
Hydrocarbons	An international agreement adopted in December 1997 in Kyoto, Japan. The Protocol sets binding emission targets for developed countries that

	would reduce the emissions on average 5.2 percent below 1990 levels.
Industrialized countries	The metric prefix for one millionth of the unit that follows.
Kyoto Protocol	An international treaty created in 1997 in Kyoto, Japan to reduce industrial nation's global emissions of greenhouse gases.
Megawatts	one million watts or one thousand kilowatts
Methane (CH_4)	A gas emitted from coal seams, natural wetlands, rice paddies, enteric fermentation (gases emitted by ruminant animals), biomass burning, anaerobic decay or organic wastes in landfill sites, gas drilling and the activities of termites.
Photovoltaics	The use of lenses or mirrors to concentrate direct solar radiation onto small areas of solar cells, or the use of flat-plate photovoltaic modules using large arrays of solar cells to convert the sun's radiation into electricity.

Energy Sustainability: Planning, Policy

1. Introduction

Energy is the driver of growth. Energy is one of the most important building block in human development, and essential factor in determining the economic development of any nation. Energy efficiency and renewable energy are said to be the twin pillars of sustainable energy.

Sustainable energy is the sustainable provision of energy that meets the needs of the present without compromising the ability of future generations to meet their needs. This Energy is replenishable within a human lifetime and causes no long-term damage to the environment

Non-renewable sources include fossil fuels, e.g. oil, coal, gas and their deposits are limited and can be exhausted. Renewable energy sources include solar, wind, biomass, hydro, geothermal and ocean power.

Greater use of renewable energy technologies will reduce dependence on fossil fuels and will bring diversity and security of supply to global energy infrastructure.

Today's excessive uses of fossil fuels is changing the world's climate. If CO_2 emissions are left unchecked, global average temperatures could be as much as 5 – 8 degrees Celsius higher by the end of this century, with devastating impacts on the economy and the natural world meeting the needs of the present without compromising the needs and opportunities of future generations.

At present we still depend a lot on fossil fuels and other kinds of non-renewable energy.

Energy is a scarce commodity and a most valuable resource. The potential of renewable energy sources is enormous as they can in principle meet many times. Energy is one of the most important resources to sustain our lives.

Fossil fuel based energy sources are facing increasing pressure on a host of environmental fronts, with perhaps the most serious challenge confronting the future use of coal being the Kyoto Protocol greenhouse gas (GHG) reduction targets.

Global climate change caused by the relentless build-up of greenhouse gases in the Earth's atmosphere is already disrupting ecosystems, resulting in about 150,000 additional deaths each year.

An average global warming of 2°C threatens millions of people with an increased risk of hunger, malaria, flooding and water shortages.

If rising temperatures are to be kept within acceptable limits then we need to significantly reduce our greenhouse gas emissions.

It is need of identify a comprehensive set of criteria and indicators for constitutes energy sustainability, then develop an agenda to achieve energy sustainability through a combination of incentives and regulatory measures for conservation and efficient use, research, and development of advanced technologies and methods.

2. Sustainable energy

Since the dawn of the industrial age, the explosive growth in economic productivity has been fueled by oil, coal, and natural gas. World energy use nearly doubled between 1975 and 2005.

China's energy use has been doubling every decade. The implications for the environment are staggering. One way or another, our reliance on fossil fuels will have to end.

Energy for Sustainability evaluates the alternatives with good planning and policy decisions, renewable energy and efficiency can support world demands at costs we can afford economically, environmentally, and socially.

Sustainable energy means energy which is used to provide the services which global need, such as heating, lighting and cooling, without contributing to global climate change or damaging the environment.

Objective of Sustainable energy is:
• Promoted renewable solutions,
• Respect the natural environment
• Point out unsustainable energy sources
• Generate greater equity in the use of resources
• Assessment economic growth from the consumption of fossil fuels

In 2002, Johannesburg, South Africa for the 10th anniversary of the first Earth Summit in Rio de Janeiro, Brazil. At the Johannesburg Earth Summit, countries committed to:
"Substantially increase the global share of renewable energy sources with the objective of increasing its contribution to total energy supply"

At the 20th Earth Summit June 2012, countries, companies, cities, and individuals need to commit to increasing the amount of wind, solar, geothermal, tidal, and wave power throughout the world to 15 percent of total electricity by 2020 more than doubling what is predicted under current trends. Civic and corporate stakeholders must commit to do more to increase electricity production from renewable sources.

Energy sustainability means the harnessing of resources that.

(1) are not depleted by continued use;

(2) do not hazards to the environment or emit toxic pollutants; and

(3) do not involve in health hazards or social inequalities.

In general renewable forms of energy are considered green because they cause little depletion of the Earth's resources, have beneficial environmental impacts, and cause negligible emissions during power generation.

Non-renewable energy sources are responsible for the greenhouse effect, causing global warming, which endangers our planet and future generations.

The world's energy demand renewable energy sources such as biomass, wind, solar, hydropower, and geothermal can provide sustainable energy services, based on the use of routinely available, indigenous resources.

A transition to renewable based energy systems is looking increasingly likely as their costs decline while the price of oil and gas continue to fluctuate.

Renewable energy supplies are of ever increasing environmental and economic importance in all countries. A wide range of renewable energy technologies are established commercially and recognized as growth industries.

3. India's energy policy

Energy is the foremost goal in India's energy policy making, as nearly one-quarter of the population lacks access to electricity. This implies ensuring the supply of adequate and reliable energy to the Indian population amid growing energy demand, bolstered by economic growth.

International studies on human development indicate that India needs much larger per capita energy consumption to provide better living conditions to its citizens.

India's second largest population and increasing pace of economic growth make its energy needs particularly challenging. India is now the eleventh largest economy in the world, fourth in terms of purchasing power. It is poised to make tremendous economic strides over the next ten years, with significant development already in the planning stages.

This report gives an overview of the renewable energies market in India. India is an emerging economy, increasing GDP is driving the demand for additional electrical energy, as well as transportation fuels.

Fossil fuel and renewable energy prices, and social and environmental costs are heading in opposite directions and the economic and policy mechanisms needed to support the widespread dissemination and sustainable markets for renewable energy systems are rapidly evolving.

India has a vast supply of renewable energy resources, and it has one of the largest programs in the world for deploying renewable energy products and systems. The extensive use of renewable energy including solar energy needs more time for technology development.

In G20 countries, Germany had the largest amount of its electricity produced from renewable sources in 2011, followed by the European Union, Italy and Indonesia. The United States ranked 7th, India came in 9th, and China ranked 12th. Spain, Portugal, Iceland, and New Zealand which each produced more than 15 percent of their electricity from these sources.

India is dedicated to the mitigation of climate change, although overcoming energy poverty and ensuring economic and social development remains a top priority.

India is suffering from huge estimated shortages of nearly 10% in energy terms and almost 17% in terms of peak demand. Integrated Indian Energy Policy (2006) mentioned as- We are energy secure when we can supply lifeline energy to all our citizens irrespective of their ability to pay for it, as well as meet their effective demand for safe and convenient energy to satisfy their various needs at competitive prices, at all times and with a prescribed confidence level considering shocks and disruptions that can be reasonably expected.

Energy security is driven by increasing dependence on imported fuels, which is crucial to meet the India's huge energy demand. Increased import dependence also exposes the country to greater geopolitical risks and international price volatility.

In India 3,700 MW are currently powered by renewable energy sources. The need of renewable energy are due to:

o Increases in population and filling of demand-supply gap

o Eco- friendly, Enormous potential

o Necessity to strengthen India's energy security

o Pressure on high-emission industry sectors from their shareholders and for rural electrification.

India's energy requirement comes mainly from five sectors; industry, agriculture, transport, services and domestic, each having considerable saving potential.

Energy needs of the country, forecasts of consumption and production, and its growth and society with renewable resources.

Human population and the individual life expectation will increase, energy could, in the future, be in short supply. Untill supply increased, source of friction in human affairs energy conservation is the deliberate practice or an attempt to save electricity, fuel oil or gas or any other combustible material, to be able to put to additional use for additional productivity without spending any additional resources or money.

The development and use of renewable energy sources can enhance diversity in energy supply markets, contribute to securing long term sustainable energy supplies, help reduce local and global atmospheric emissions, and provide commercially attractive options to meet specific energy service needs in developing countries and rural areas helping to create new employment opportunities.

To better understand the current situation in India and the future of the renewable energies market look at the trends in energy consumption, growth of the current grid and the availability of transportation and equipment used.

The development and deployment of renewable energy, products, and services in India is driven by the need to decrease dependence on energy imports sustain accelerated deployment of renewable energy system and devices expand cost effective energy supply.

Any effort to maintain atmospheric levels of CO_2 below even 550 ppm cannot be based fundamentally on an oil and coal-powered global economy, barring radical carbon sequestration efforts.

The sustainable energy management system is a far-reaching concept that covers all aspects of the sector from fuels and their extraction to energy generation and systems efficiencies, energy distribution, and energy consumption, and energy security implications etc.

These three ministries look after the entire energy policy of India. Electricity in India is a concurrent subject at entry 38 in list III of the 7th schedule of the constitution of India. This implies that both the Union and the States have power to legislate on matters relating to electricity.

Renewable energy sources account for 31% of the India's primary energy demand. Biomass, which is mostly used in the heat sector, is the main renewable energy source. The share of renewable energies for electricity generation is 15,5%. The contribution of renewables to heat supply is around 63%, to a large extent accounted for by traditional uses such as collected firewood. About 68% of the primary energy supply today still comes from fossil fuels. The electricity sector will be the pioneer of renewable energy utilisation. By 2050, around 69%of electricity will be produced from renewable energy sources. A capacity of 1,659 GW will produce 3,860 TWh/a renewable electricity in 2050.

Within the electricity act 2003, it requires each State regulatory commission to specify the minimum percentage of electricity that each distribution utility must source from renewable energy sources. In Energy management, environmental, economic and social results of fuel extraction and transportation from its source to its destination; the energy requirements and resource intensity needed to manufacture the choices related to energy use have a strong impact on environmental sustainability.

The National Electricity Policy 2005 stipulates that progressively the share of electricity from non-conventional sources would need to be increased; such purchase by distribution companies shall be through competitive bidding process; considering the fact that it will take some time before non-conventional technologies compete, in terms of cost, with conventional sources, the commission may determine an appropriate deferential in prices to promote these technologies.

In the last three goal that India pursues viz. energy access, energy security and mitigation of climate change. These three are closely related, but sometimes conflict with one another and are derived from the reality in India. Thus, it is challenging for India to maintain a balanced approach in pursuit of all three objectives.

4. Energy efficiency

Renewable energy and energy efficiency are sometimes said to be the "twin pillars" of sustainable energy policy. Both resources must be developed in order to stabilize and reduce carbon dioxide emissions. Efficiency slows down energy demand growth so that rising clean energy supplies can make deep cuts in fossil fuel use.

If energy use grows too fast, renewable energy development will chase a receding target. Thus, given the thermodynamic and practical limits of energy efficiency improvements, slowing the growth in energy demand is essential. However, unless clean energy supplies come online rapidly, slowing demand growth will only begin to reduce total emissions; reducing the carbon content of energy sources is also needed. Any serious vision of a sustainable energy economy thus requires commitments to both renewables and efficiency.

Climate change concerns coupled with high oil prices and increasing government support are driving increasing rates of investment in the sustainable energy industries, according to a trend analysis from the United Nations Environment Programme.

According to UNEP, global investment in sustainable energy in 2007 was higher than previous levels, with $148 billion of new money raised in 2007, an increase of 60% over 2006. Total financial transactions in sustainable energy, including acquisition activity was $204 billion. This can play important roles in furthering energy conservation, efficiency, and renewable objectives, even though they do not have the land use authority of cities or counties.

5. Renewable energy potential

Global futures for renewable energy resources is booming. Decades of technical progress have seen renewable energy technologies such as wind turbines, solar photovoltaic panels, biomass power plants, solar thermal collectors and many others move steadily into the mainstream.

The global market for renewable energy is growing dramatically; in 2007 its turnover was over $70 billion, almost twice as high as the previous year. The time window for making the shift from fossil fuels to renewable energy, however, is still relatively short.

A decision taken to construct a coal or gas power plant today will result in the production of CO_2 emissions and dependency on the resource and its future costs lasting until 2050.

The future of renewable energy development will strongly depend on political choices made by both individual governments and the international community. At the same time strict technical standards will ensure that only the most efficient fridges, heating systems, computers and vehicles will be on sale.

6 . References

1. Energy for a sustainable world: Jose Goldenberg, Thomas Johansson, A. K. N. Reddy, Robert Williams (Wiley Eastern).
2. Modeling approach to long term demand and energy implication : J. K. Parikh.
3. Energy Policy and Planning : B. Bukhootsow.
4. World Energy Resources : Charles E. Brown, Springer2002.
5. 'International Energy Outlook' - EIA annual Publication
6. Principles of Energy Conversion: A.W. Culp (McGraw Hill International edition.)
7. Aspects of Energy Conversion : I. M. Blair and B. O. Jones
8. Principles of Energy Conversion : A. W. Culp (Mc Graw Hill International)
9. Energy conversion principles : Begamudre , Rakoshdas
10. Principles of Energy Conversion : A.W. Culp.
11. Energy Management: W. R. Murphy, G. Mckay (Butterworths).
12. Energy Management Principles: C. B. Smith (Pergamon Press).
13. Efficient Use of Energy : I. G. C. Dryden (Butterworth Scientific)
14. Energy Economics -A. V. Desai (Wieley Eastern)
15. Industrial Energy Conservation : D.A. Reay (Pergammon Press)

16. Energy Management Handbook – W. C. Turner (John Wiley and Sons, A Wiley Interscience Publication)
17. Efficient Use of Energy: I. G. C. Dryden (Butterworth Scientific)
18. Energy Management Handbook – W.C. Turner (John Wiley and Sons, A Wiley Interscience publication)
19. Industrial Energy Management and Utilisation –L.C. Witte, P.S. Schmidt, D.R. Brown (Hemisphere Publication, Washington, 1988)
20. http://learn-energy.net/education/renewables.php
21. The European Geothermal Energy Council (EGEC), http://egec.info/publications/
22. http://www.eu-oea.com/technology-2/
23. http://energyquest.ca.gov/story/index.html

ABOUT THE AUTHOR

Dr. Hemant Pathak held positions as Assistant Professor in the department of chemistry, Govt. Indira Gandhi Engineering College, Sagar, MP, India. He had extensive experience in teaching, research and administrative management.

Dr. Pathak received his Ph.D. degree in chemistry from Dr. Hari Singh Gour Central University, Sagar, India and M.Sc. Gold medalist from Jiwaji University, Gwalior. He has published 18 books and more than 50 research papers in reputed International and National journals and received several awards. He is a member of editorial boards and reviewer boards of several international journals and societies. His area of specialization includes Engineering Chemistry, Energy audits and Environmental Pollution management.

www.ingramcontent.com/pod-product-compliance
Lightning Source LLC
Chambersburg PA
CBHW071600170526
45166CB00004B/1737